Alvan Edmond Small

Causes that Operate to Produce the Premature Decline of Manhood,

and the best means of obviating their effects and bringing about a

restoration of health

Alvan Edmond Small

Causes that Operate to Produce the Premature Decline of Manhood,
*and the best means of obviating their effects and bringing about a restoration of
health*

ISBN/EAN: 9783337373436

Printed in Europe, USA, Canada, Australia, Japan

Cover: Foto ©berggeist007 / pixelio.de

More available books at **www.hansebooks.com**

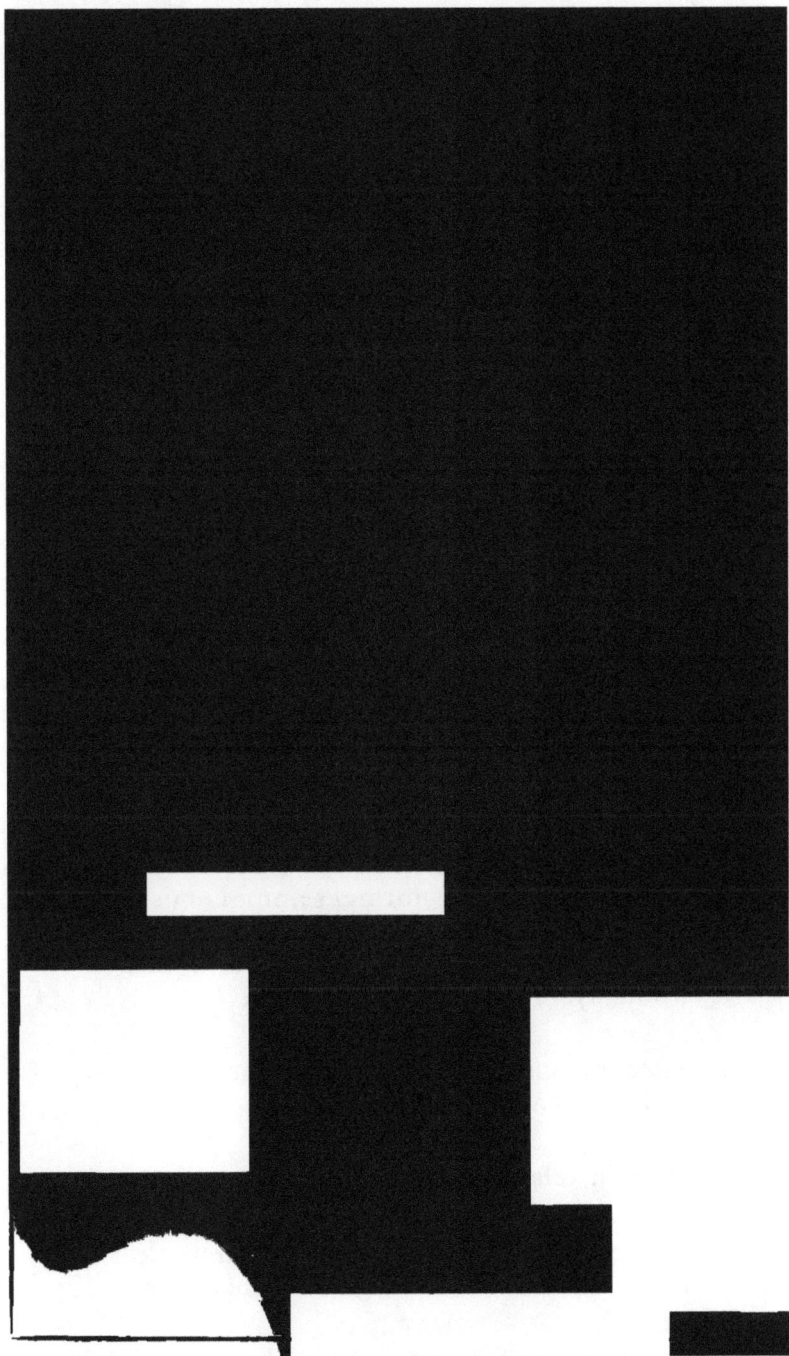

TABLE OF CONTENTS.

PREFACE.

In furnishing the following treatise on the " CAUSES THAT INDUCE THE PREMATURE DECLINE OF MAN-HOOD," and the most judicious means of removing them and curing their effects, the author is aware of the many difficulties in the way of producing anything like a satisfactory treatise upon the subject. But from many years experience he has collected the results of his observations into the following pages.

The remedies employed have been such as he has found most effectual, and yet he is aware that in the hands of many practitioners other remedies have been employed not mentioned in this treatise.

The work is not arranged so methodically as the author could have wished, and yet his object will be accomplished if he has made any suggestions or thrown any light upon the subject that will aid the profession or redound to the benefit of the unfortunate.

With the hope, therefore, that this attempt to supply a want in the literature of Homœopathy will be charitably received, and pave the way for better efforts, the work is given to the profession and public.

INTRODUCTION.

The chief object to be accomplished in writing a treatise on the causes that produce a premature decline of manhood, is to point out, explicitly, the best protection against them, as well as the most commendable measures of relief from their effects.

2. Sexual debility results from a variety of causes, which, in many instances, are avoidable, while in some instances it may be constitutional and dependent upon a general deterioration of the nutritive and nervous systems.

3. When the cause is known the first inquiry is, can it be removed? and can the effect produced be remedied? No one suffers from the malady with indifference, and, therefore, it may be concluded that every victim to sexual disorder desires radical relief.

4. From the nature of the affection, those who suffer are prone to seek aid from any source that promises it, and, without some information of a specific nature, that will lead to a proper discrimination, a resort to nostrums and quackery, injurious in their results, too frequently happens.

5. As there are no two constitutions alike, and seldom that two cases of suffering are so precisely alike that the same treatment will suffice for each, it is proposed to point out the several grades of the disorder in connection with the causes and treatment of each, when medical treatment is required. The kind of debility particularly under consideration in the following pages is generally known under the head of "Involuntary emissions of semen," which operates disastrously upon the vital condition of manhood.

CHAPTER I.

A GENERAL VIEW OF THE SUBJECT.

One of the main characteristics of sexual debility is frequent involuntary emissions of the seminal fluid, followed by a feeling of exhaustion. For strong, healthy and plethoric persons these emissions may occur occasionally without exciting undue apprehension of the result. But even in such persons, where the occurrence is frequent and copious, great weakness and lassitude is apt to follow, betokening an abnormal condition, that requires both hygienic and medical treatment.

2. In those of vigorous manhood, and of superabundant vitality, the involuntary evacuation of the seminal vessels at times may be considered a conservator of health, and demanded for its protection. This is true so long as such persons remain conscious of their occurrence, unaccompanied by general malaise and lassitude.

3. Lascivious dreams are simply reflex symptoms conveyed from the distended seminal vesicles through the nerves of the brain ; and the most salutary remedy for emissions arising from this cause is through matrimonial alliance and the legitimate exercise of the function.

4. But emissions that occur without the cognizance of the patient, and without erections and pleasurable sensations, are of a different character, and may result from an abnormal irritability of the sexual organs, and, if not arrested, they will deteriorate the general health and produce a gradual decline of virility.

5. The causes that operate to produce increased irritability are various, and merit separate consideration in order to arrive at the means of obviating them, and the most rational application of remedies to overcome the debility and promote strength. Some of these causes are avoidable, and require but a strong effort of the will to set them aside ; others are unavoidable, being

diseased conditions of the urinary organs, from which the increased irritability of the sexual organs proceeds. These causes can only be removed by well-chosen remedies, or by the skill of the operative surgeon.

6. From either class of causes excessive emissions may occur, differing only from normal seminal emissions in being involuntary and debilitating. The seminal vesicles, ducts, and their orifices may severally participate in the irritability, and become so debilitated that seminal discharges are found to occur frequently, and on the slightest provocation. The simple act of urination, the effort at stool, or immediately after both, or when brought into proximity with female society, severally operate to provoke seminal emissions, without erections or pleasurable sensations, or consciousness of the loss. This is properly termed "spermatorrhœa."

7. Pathologically considered, these seminal evacuations arise from a variety of causes, and not, as is frequently asserted, from sexual excesses and abuses. They often result from other and quite different causes.

8. When seminal emissions arise from causes inherent in the system, their origin is so wrapped in obscurity that the real cause is overlooked, and, consequently, they are found to be the most persistent and difficult of cure. It is only when the true pathology of the disorder is discovered that we can arrive at a knowledge of the starting-point, so as to institute a rational or successful treatment. Anything short of this necessarily subjects the patient to a continual drain upon his constitution and life. For without duly referring the symptoms to the veritable source from whence they proceed, so as to base a proper and curative treatment, in spite of remedies given in accordance with symptoms alone, the difficulty will progress until the health is undermined and the virility destroyed.

9. The matter must be probed to the bottom, or otherwise the symptoms are meaningless, and afford but an empty display of phenomena, as likely to mislead in the treatment as to favor its utility ; for as a variety of causes, differing widely in their nature, may

produce like effect, genuine skill, in the treatment of spermatorrhœa, must depend upon a knowledge of the real cause as a starting-point.

10. A glance at the anatomy of the genito-urinary apparatus and its relation to neighboring parts, will convince any rational practitioner of the reasonableness of this assertion ; for who can deny the fact that hard impacted fæces, pressing upon the seminal vesicles, operates to produce more or less disorder in the sexual organs? Who does not know that a mismanaged gonorrhœa produces the most disastrous consequences upon the testes? And why? Because of their intimate anatomical connexion.

11. Therefore, let us look after the various causes of seminal weakness, which, as so many fountains, send forth abnormal excitement to the genital organs, producing emissions and spermatorrhœa.

1. Urethritis from different sources.
2. Stricture of the urethra.
3. Affections of the rectum and anus.
4. Constipation, retained fæces.
5. Hemorrhoids and varices or fissures.
6. Intestinal worms.
7. Chronic inflammation and tenesmus of the bladder.
8. Stone or gravel.
9. Spinal irritation.
10. Sexual excesses.
11. Masturbation.
12. Prostatitis.
13. Morbid imagination.

A careful examination of each of the above will enable us to rationally comprehend the relation between cause and effect in each particular case, and show us the total insufficiency of a merely symptomatic treatment of these diseases.

12. Either one or all of these causes have a close relation to the genital organs, and a continual irritation from either forms a source of determination of blood, congestion, swelling, excessive irritability of the

testes, spermatic cord, seminal vesicles, prostate gland, and veru-montanum, and thence erections at first, and seminal emissions. In time the erections cease; the membrum-virile becomes flaccid, and the fluid constantly passes off, altered in quality, thin, and void of the characteristic spermatozoa. With this general view of the subject we may form some idea of the nature and extent of the malady under consideration.

CHAPTER II.

SPERMATORRHŒA FROM INFLAMMATION AND IRRITATION OF THE URETHRA.

Chronic inflammation of the urethra, from whatever cause, extends itself by means of continuity of structure not only to the bladder, uterus and kidneys, but also to the organs of reproduction. This inflammation may be produced, 1st, By mechanical injury, inflicted by instruments used by onanists, such as pencils or quills, which they introduce into the urethra to excite pollutions, when, from frequent and long continued abuse, the urethra becomes insensible to other excitants. Fragments of such instruments have been known to remain, forming a nucelus for inflammation.

Blows, bruises, and other occasional injuries may provoke spermatorrhœa. The great sensitiveness of the urethra, and the unavoidable admixture of the urine with the blood, from such injuries, renders wounds or mechanical abrasions or squeezes of the greatest importance.

To cure a spermatorrhœa, originating from causes of this kind, the unnatural use of the instruments must be banished; all pressure must be removed, and all positions leading to contusion or pressure upon the organ should be avoided, and then let moderately cold water be freely used for ablutions twice or thrice during the day. *Arnica* at first is an excellent remedy for internal administration, and drop doses may be taken several times daily until the soreness and irritability of the parts disappear; when the cause is removed the effect

ceases. Should *arnica* prove insufficient, injections of
tepid water, with a few drops of the tincture of *calen-
dula*, may suffice. The distilled extract of *hamamelis*
may be used for the same purpose.

2. Certain drinks and articles of diet produce an unfavorable action upon the urethra, and sometimes cause
inflammation. Certain medicines, taken in excess,
have the same influence, such as diuretics. The irritation produced by some liquors, such as gin, are proximate causes of involuntary seminal emissions. When
oysters, crabs, or other shell-fish exert an injurious action of the kind, refrain from using them. The same
advice is given with reference to articles of food or
drink or medicine known to exert an unfavorable effect
upon the genital organs, and the spermatorrhœa that remains as an effect can be easily cured by well-chosen
remedies.

A young clerk in a silk house suffered from involuntary
emission several times a week, until the effect upon his
health began to be noted. He applied to a physician,
who inquired critically into his habits, all of which he
found quite regular, excepting eating late at night. On
inquiry it was found that he indulged freely in such
food as eggs and oysters. He was advised to confine
himself to regular meals, and to avoid both the eggs
and oysters—to partake of beef, mutton, and vegetables,
and avoid poultry and stimulating soups. At first but
little change was noted ; his digestion seemed somewhat
impaired. *Nux vom.* 3d was given half an hour after
each meal, and he soon recovered from the emissions
and gained robust health.

Another young man, who suffered much from this
difficulty, was constantly annoyed by difficult digestion
and depressed spirits, was cured by *pulsatilla*, 3 ; dose
three times a day, after duly regulating his diet.

3. Another source of inflammation of the urethra is
infection from impure sexual exposure and in urethral
stricture from chancre ; from these an irritation is set up
which results in pollutions of a serious character. To
cure these, if they result from simple urethritis or gon-

orrhœa, and particularly if there is much smarting when urinating, *aconite* may be taken first, and afterwards *cannabis sativa*. Several cases of this kind have been noted: A gentleman of middle age having contracted gonorrhœa, which he supposed himself entirely cured of, was afterwards the victim of nocturnal emissions, which weakened his whole system. He was greatly depressed, and suffered inveterate constipation. He had applied to one physician after another without benefit, until he despaired of finding relief. He finally consulted a Homœopathic physician, who minutely inquired into his antecedents, and then prescribed *nux vom.* to remove the constipation; after *nux lycopodium* was given, and his bowels moved freely, without difficulty. The 6th attenuation of these remedies was used. A diet of toast and steak in the morning, with a cup of black tea; beef or mutton with vegetables for dinner, and brown bread or toast for tea, was directed, *cannabis sativa* 6th was given every six hours, for a week, in connection with this diet, and the patient found himself greatly relieved; and this treatment was afterward continued until he was completely rid of the difficulty.

Other cases were cured by addressing remedies directly to the proximate cause, and afterwards to the resulting symptoms. When the urethral irritation is subdued, and the pollutions still continue, *cannabis sativa* or *cantharis* will generally cure.

When stricture of the urethra is ascertained to be the existing cause of emissions, the stricture must be removed before the difficulty can cease. Sulphur 3d will often accomplish the whole, in connection with a well regulated diet. To secure regularity of the bowels, *nux v., sulphur* and *mercurius sol.* may be used advantageously.

CHAPTER III.

MASTURBATION.

We shall now consider a cause of urethral irritation which by many is believed to be one of the most deteriorating vices incident to the youth of both sexes. Very early in life boys have been initiated into habits of self-abuse, and if no one finds an opportunity to advise and counsel them otherwise, by pointing out the dangerous consequences of the practice, the habit becomes fixed and difficult to break. After a while such a youth begins to grow sickly, pale, nervous, and unmanly, and unless the habit is arrested in time, and the victims become sensible of the danger, his approach to manhood is fraught with entailed consequences, that finally render him an object of pity and disgust. Youth of fifteen or sixteen summers, who indulge in this vice, may soon bring about an irritation of the urethra that merges into real inflammation, attended by a secretion which has a close relation to pollutions and spermatorrhœa. It is our purpose, therefore, to treat this subject somewhat in detail, in order to give an insight into the best remedial measures for the protection of virility.

2. Immediately before the age of puberty and during the transition period, as well as after, unknown longings and desires assert their supremacy in the youthful mind. These are the first utterings of sexual instinct, though scarcely recognized as such by the subjects. Some arrive at this period earlier and some later. In peculiarly irritable constitutions the instinct after puberty manifests itself with great intensity, and exhibits its effects in certain tendencies to bodily ailments and frailties.

3. When by the allurements of others such youth are initiated into solitary habits of exciting the genitals, first by filling the mind with impure thoughts and imaginings, and then by manual interference, the most pernicious and degrading habit of self-pollution steals over them, and soon results in physical weaknesses which are prophetic of the early decline of manhood.

4. The defect of early education, coupled with direct temptation and the allurements of evil association, often form the basis or starting point for a career of miserable habits, that sooner or later prey upon the mental and physical constitution. When once initia-

ted and predisposed. both mind and body become subject to excitement from otherwise trivial causes. Novel reading, love stories and sickly sentimentalism of every kind, become disastrous allurements to the habit of masturbation. These allurements and dangers often override the rigid supervision of parents and educators, and beset the youthful and impressible mind in many different ways.

It is a fact no less sad than true, that very many in their tender years, and even in childhood, are brought through one means or another into mental excitement and thence into actual abuse of the genital organs, which becomes a habit of greater or less injury.

By such abuses the genital organs become impaired and diseased, and the strength and life of the entire organism becomes undermined.

These excesses are so extremely dangerous and inevitably ruinous to body and mind that it is of the utmost importance to detect and arrest them before it is too late, and to promote this end young boys should be subject to the kindest and most friendly supervision, and every prudent effort should be made directly and indirectly to withdraw them from such influences and habits as excite sexual phantasies, and without acquainting them with the friendly object intended, great care should be exercised in diverting their minds from all misleading thoughts and inclinations before they have become so rooted that all attempts to obliterate them prove a failure.

To accomplish this requires:

1st. A provision of suitable and useful employment for mind and body.

2d. To withhold all kinds of seductive reading and theatricals, and all imprudent mixing with the opposite sex, such as allowing them after early childhood to occupy the same sleeping apartments and to share the same couch. Experience and observation have established beyond a doubt the imprudence, if not the wickedness, of allowing such practices.

3d. But at suitable times and under the supervision

of proper restraints, boys and girls should be encouraged in outdoor sports and open air exercise. It is a mistaken idea to suppose that the cause of virtue is promoted by keeping them secluded from the society of each other. For such is not the case. The reverse often begets that sickly, unhealthy secret brooding that creates a want of self-respect and dread of social intercourse. It is far better that boys should have the advantage of associating in suitable plays and pastimes with girls of their own age, while at the same time every moral and elevating influence should be exercised over them by their parents and guardians.

4th. If a suspicion exists that boys indulge in self-abuse, let them be quietly watched until the truth is ascertained. An inclination to be alone and in secluded places often excites suspicion of self-abuse, and especially if after such seclusions they seem pale and excited, or depressed and morose or peevish and fretful. And unusually timid boys and young men addicted to self-abuse are generally shy and timid. They betray an absence of manhood and appear cowardly, and moreover the countenance betrays with nearly the same certainty as it does the tippler or opium eater, and when interested and affectionate friends learn from these signs the painful fact, their friendly aid is demanded, they should point out to the erring youth the sad effects that must inevitably follow this habit. Many a boy, bright and lively at ten years of age, has become dull, sickly and mentally dejected before his twenty summers by such miserable self-abuse. A slow, hesitating speech, a blundering mode of speaking, a bad memory and dullness of apprehension in general, are the usual fruits of this sin ; and since so much is at stake, and the guilty ones are so prone to deny their pernicious habits, no prudent means of detection should be neglected. If necessary the bed linen and shirts should be called into requisition to afford confirmation of the fact, and if caught in the practice let them be told in a fatherly and affectionate way the disastrous consequences, not once only, but many, many times, until the habit is subdued

5th. This course will hardly fail to make a favorable impression on young and pliable minds, and if the will power at first is inadequate, gentle encouragement and the aid of solicitous parents will greatly stimulate resolution till the habit is conquered.

When made fully sensible of the dreadful consequences of self-abuse, there are but few so reckless and incorrigible as to countenance and continue the habit. Mechanical means of restraint has sometimes been found necessary to aid the efforts of young men in dissipating the inclination and practice. The best medical counsel may also be necessary to aid in bringing about this desirable result, and it would be adding crime to crime to withhold any measure capable of arresting the criminal practice.

We have thus far treated of masturbation as a crime involving the voluntary surrender of manhood. But now we propose to consider other causes that may induce the habit, that places the victim more in the light of a sufferer than in that of a sinner. Ascarides or pin worms in the rectum sometimes induce an itching that implicates in no small degree the genitals of young boys, and compels them to scratch and rub until they unconsciously fall into the habit of masturbation. Herpetic eruptions that burn and itch may compel similar habits, terminating in the same way. The reason why the habit operates so injuriously upon the entire organism, whether induced by vicious allurements, morbid broodings, or sickly love stories, or by diseased conditions as detailed above, is because of the close proximity and relation of the sexual apparatus to the spinal cord, thence to the brain, which first receive the shock and then by reflex action the urinary and digestive systems become affected. But the injury does not stop here ; the mind becomes filled with impure images exciting the brain, and thence the memory and mental faculties. It finally ends in hypochondria, melancholy and suicide, or in epilepsy, apoplexy or dementia.

The intensity of the shock which the habit imparts to the whole system, and the consequent prostration and debility must especially prove injurious to the organs properly supplied with nerves. The stomach, therefore, is liable to become distended, and nutrition is seriously interrupted ; the entire body as a consequence fails of receiving sufficient nourishment, it gradually gets weaker, the nerves become unstrung and disposed to paralytic weakness and cramps ; and as during the act there is a determination of blood to the brain, frequent cases of apoplexy have terminated the fatal habit ; or otherwise ten or fifteen per cent. of the number of inmates of the lunatic asylums prove to have been the victims of this vice.

Self-abuse has another peculiar effect, if not early arrested : it destroys sexual enjoyment and entails coldness in the conjugal bed, and begets disappointment and unhappiness and completely frustrates the fulfilment of marriage relations.

It is to be feared that many who attain to a marriageable age refrain from entering into wedlock by reason of the cold indifference toward the institution, which results from this vice. They prefer the mistaken enjoyment of solitary self-abuse to the satisfaction of legitimate sexual intercourse.

It will be seen, then, that self abuse requires greater exertion of the sexual organs than a natural embrace, and that they are weakened by oft-repeated and unnatural irritation, as the opportunities are more frequent than for natural intercourse, and the constrained position of standing or sitting must prove more injurious to muscles and nerves that are brought into requisition, it follows conclusively that greater muscular and nervous weakness must result from the act. It is also manifest that in a natural embrace the excitement is participated in by two, whereas in self-abuse it is entirely cultivated by one's own imagination, and must therefore prove the more exhausting. Self-abuse is more dangerous because it withdraws its victims from society and leads to solitary dwelling upon themselves, or to musings and broodings of so low a quality that they fall into the pit of melancholy and thus make complete shipwreck of health and life.

In order to place safeguards around unsophisticated youth and boys in early life, parents and teachers must be vigilant and ready to comprehend the importance of directing their thoughts and habits in the right direction at first. They should teach them to avoid such plays as suggest and often prove the beginning of injurious habits. It is far better to avoid the causes of urethritis at first than to be successful even in curing pollutions and spermatorrhœa.

As boyhood and youth passes away, and manhood supervenes, strictures produced by indiscretions of youth, may be the constant cause of spermatorrhœa. These strictures arise from various causes ; the worst and most difficult to cure are for the most part traced to badly treated gonorrhœa. But those arising from the cicatrices of wounds, or from urethral chancre, are caused from the folding of the mucous lining of the canal, and are sufficiently formidable. For the folding and swelling and formation of valves by fibrous indurations of the mucous membrane, are tedious in duration and difficult to cure. Strictures generally have their seat in that portion of the urethra where the inflammation has been the greatest and the suppuration the most apparent. There may be more than one near or more remote from the orifice of the bladder. The narrower and longer they are the greater the obstruction to the passage of urine and semen, and the portion posterior to the strictures becomes distended, and presses indirectly upon the seminal vesicles so violently as to weaken and injure, and thus be the source of an obstructed spermatorrhœa.

Strictures may also produce other obstructions more and more interiorly until effete and irritating matters which should be thrown off meet with obstructions, and are thrown back upon the delicate seminal vesicles, and result in weakening them to a degree that destroys their normal functions, or so paralyzes the tenacity of the ducts as to allow the seminal fluid to flow off unnoticed without any sensation or erection ; in this way spermatorrhœa results, and impotence becomes inevitable.

The effects of masturbation not only extend to the urethra, prostate gland and bladder, but to contiguous structures. The rectum becomes affected—the sphincter-ani contracts, varices and hemorrhoids are severally the result of urethral irritation. Stricture and frequent urging to urinate, hemorrhages from the bowels, prolapsus and all the various hemorrhoidal ailments sometimes refer themselves to the same cause, though too frequently overlooked, and on this account a misdirected treatment is liable.

Spasmodic strictures as well as those caused by urethral inflammation, may result from self-abuse, and also those resulting from enlarged veins of the urethra, from gouty deposits, urinary calculi, and forcible catherization, may provoke spermatorrhœa or emissions.

Since strictures are so often the cause of seminal weakness, and self-abuse may be the first cause of the urethral inflammation that produces them, it is well to be familiar with the symptoms of stricture that they may receive early attention. The first symptom is the retarded flow of urine, and also the urging to force the urine by reason of some sensible obstruction, behind which the urine accumulates ; the stream is changed in form and volume, at first very small, afterward sluggish, and finally a mere dribble from the urethral orifice. This departure from the normal standard is so marked that no one can fail to discover it.

CHAPTER IV.

HOW TO CORRECT THE HABIT OF SELF—ABUSE AND CURE ITS EFFECTS.

1. It is for parents and guardians to impress extremely young subjects with the wickedness and danger of the practice, to watch over them and correct them for every known indulgence. There is but little difficulty in correcting this abuse while the victims are very young and before the habit is confirmed.

2. When boys are associated together and become each other's instructors in the vice, let such associations

be broken up. Older boys often initiate younger ones
and fill their susceptible and tender minds with lewd
thoughts, and great care to guard against evil associations is absolutely requisite when such, before the age
of puberty, begin to exhibit that peculiar cast of the
countenance and debility consequent upon such early
vice. Let them be impressed that the debility and
sickness is brought on by the crime. Strengthen the
impression by holding up the terrible and almost fatal
results that will surely follow, if they habitually re-
peat the act or allow themselves to be misled by
wicked associates. When first debilitated, give such,
in connection with moral restraint, *cinchona*, 6th dilution, ten drops in half a tumbler of water, and give a
desert spoonful three times a day. The above is the
proper treatment when the self-abuse has been wicked-
ly initiated by older delinquents.

3. Parents should so carefully guard the health of
their children as to obviate all diseased conditions liable to induce scratching or rubbing the
genitals, for in this way the habit of masturbation
may be acquired; pin-worms or seat-worms of the anus
produce a disastrous itching, and so do herpetic eruptions in the vicinity of the genitals, and too great care
cannot be exercised in determining whether or not
such disease exists. If the fact becomes established:
for the pin-worms, give *sulphur*, 6th, ten drops in half
a tumbler of water, a dessert spoonful every night; if
this does not allay the itching, follow with *nux vomica*
in the same way, and teach the boy not to scratch or
rub. If there is any eruption that causes much itching
of the parts, *sulphur* taken as above is required. In
scrofulous children these troubles are liable to occur
and be the means of spontaneously initiating them into
the habit of self-abuse; cure the cause, and with
proper instruction the effect will cease. *Ammonia* 6th,
calcarea carb., 6th, and sometimes *petroleum*, 6th, with
suitable moral restraint, will prove sufficient to correct
these abuses.

4. To cure self-abuse in boys after they have passed

the age of puberty, requires in part the same measures as for those younger. They must be impressed with the heinousness of the crime, if voluntarily committed ; and also with the inevitable consequences upon the physical and mental health. Nothing can cure them or break them but a determined resolution to renounce the habit. To aid them in doing it, suitable employ-ment for mind and body must be provided ; and they must be made to realize that the practice is a sin against God. No protection against the vice is certain but a voluntary refraining from it from the highest of all motives ; that it is a sin against God and an abuse of themselves.

5. To correct these morbid states of mind that favor the habits, let the youth be supplied with entertaining and useful reading, and if already there is more or less disturbance of the organic functions, which react upon the brooding and susceptible mind, remedial measures are necessary.

6. To guard against the disposition to self-abuse, *calcarea carb.*, 3d trituration, may be given every even-ing for a week. If there is any disturbance of the digestive system follow the *cal. c.* with *nux vomica*, 3d, every evening one hour before retiring for another week, observing at the same time to engage in such amusements as chess and other intellectual games, and sometimes in dancing or base ball. If the onanist finds himself weak, let him take *china*, 3d. If he has stiff-ness of the back or some pain in the small of the back, let him take *cocculus*, 3d. If after the habit is sub-dued any debility remains, *china*, 3d, may be taken three times a day until the debility is overcome. When the genitals are easily excited and suggest a return of the habit, *phos. ac.*, 3d, in water will be found useful. Morbid erections and lascivious dreams require *can-tharis*, 3d, three times a day.

7. We have thus pointed out the ways that generally lead to self-abuse, as well as the surest means of ar-resting the habit, and it may be remarked that onan-ism once fixed is classed among the most inveterate of

habits. But it can be cured if proper attention is bestowed to the right kind of discipline and medication. The longer the habit remains the more deteriorating the effects upon the nervous system, and thence upon the nutritive functions. For the acute effects which arise from excessive rather than from long continued abuse we have named as remedies, *china, cocculus* and *phos. ac.*, to which may be added *merc. sol.*, 3d, and *phosphorus*. But for slow chronic effects, *sulphur* may be given at long intervals, or perhaps *carbo vegetabilis*, and a generous diet.

For weakness and flaccidity of the penis, *china* may be taken three or four times a week.

For excessive nervousness and timidity, *calc. phos.* may be taken in daily doses.

If there is a dull, indefinable bewilderment and subacute headache, *nux vomica*, or perhaps *zincum met.* will often cure.

For the minor consequences that remain after the habit has ceased, the above remedies are of the greatest importance.

CHAPTER V.

CONTINUATION OF THE EFFECTS OF SELF-ABUSE, NOCTURNAL EMISSIONS.

By these emissions are understood seminal discharges that occur during sleep, and are dependent upon an enfeebled condition of the seminal vesicles, and irritation as the predisposing cause, that for the most part result from previous self-abuse. They differ from spermatorrhœa in taking place from excitement of the genital organs and erections, which seem to be produced by lascivious dreams, or by reason of the bladder being filled with urine. The quality of these discharges does not differ materially from the normal character of healthy semen ; nevertheless they result from debility that has been induced from some cause of irritation of the urethra or weakening of the seminal ducts.

2. Those which are entailed as an effect of masturbation first demand our attention, because they occur frequently and sometimes nightly, and are followed by great weakness and depression. The victims of these emissions often are made unhappy and wretched, because they know that habits which they have succeeded in conquering were the primary cause, and they often seem willing to endure the suffering than to apply for relief. But disease from any cause may be curable, and such is the case with regard to these nocturnal emissions. We have had considerable experience in the treatment of such cases, and undoubtedly they are generally curable.

3. In the case of a young man who had been guilty of self-abuse until seventeen years, on being aroused to his condition he voluntarily abandoned the habit, and turned his attention to a conscientious religious life. But his whole system was greatly impaired. He was dyspeptic and nervous, depressed and melancholic, and found himself the victim of almost nightly emissions. He sought a confidential medical adviser, who at once comprehended the nature of his suffering For more than a year he had refrained from the polluting habit, hoping that his health and strength would return to him without obliging him to resort to a physician. But on the contrary, he found his health declining he fancied himself the victim of imbecility, and his mind was filled with fearful forebodings of a disastrous future.

4. The first measure of relief resorted to by his physician was to encourage his hopes and direct his mind from himself to the consideration of topics that would be most likely to give a healthy tone to his moral and religious aspirations. He further taught him that the study of arithmetic and the solution of its problems upon slate or blackboard would strengthen his manly faculties, whereas the reading of sentimental stories would have the opposite effect. All of which the young man seemed to appreciate.

5. After pointing out a mental and moral course that

would favor a radical cure, the physician commenced with him a course of medical treatment. Finding his patient the victim of excessive sexual excitement, he first gave him the sixth decimal attenuation of *cantharis*, four globules three times a day, and prohibited the use of all stimulating drinks, including coffee and all stimulating aliments calculated to excite sensual feelings, such as oysters, crabs, lobsters, etc., etc. This first prescription was continued for five or six days, and the young man felt a sensible relief. But feeling dull and stupid, and having a dread of society and still suffering, but less frequently, from emissions, *phosphoric acid*, 6th, was given in water three times a day ; there was a gradual improvement, and the emissions were less frequent. The patient, as directed, confined himself to a light nutritious diet, under which his strength improved ; occasional seasons of malaise and weakness were cured by *china*, 3d. For two years his whole system became more and more robust, and he rejoiced in finding himself radically cured.

6. For the debility brought on by onanism, *china* is quite generally the best remedy, and it only requires an early cessation of the habit, a careful diet, and a good occupation for the mind, as well as a persistent perseverance with this remedy to effect a perfect restoration to health.

7. For emissions brought on exclusively by this vice, attended with extreme sense of weakness, provided a proper attention is paid to diet and employment for the mind, *china* will seldom fail of curing.

8. For the depression of spirits and nervous restlessness, *hypophosphite of lime* in the third trituration, taken in 3 grain doses morning and evening, will suffice. But the remedy must be taken a sufficient length of time to ensure its effect.

9. Nightly emissions have been cured by *digitalis* in those of bilious temperament subject to melancholy. The 3d dilution is employed, but for any treatment to be successful the cause must be removed The diet should be well regulated, the mind must be accustomed

to dwell upon profitable subjects, and there must be a firm reliance upon the remedies. In addition to those already cited for nocturnal emissions, *sepia* often has a salutary effect, and in some inveterate cases *sulphur*, 3d, taken in daily doses, will effect a radical cure.

10. A student of theology, aged 25, found his health declining on account of the debilitating effects of nocturnal emissions, which were brought on by self-abuse in early life. He applied for medical treatment. The 3d aqueous dilution of *phosphoric acid*, while under a strict regimen as to diet, was administered three times a day for a month ; after which his health improved rapidly, and he rejoiced in finding himself no longer troubled with the emissions. He was enabled to pursue his studies without difficulty.

11. A young man desirous of entering into matrimony, hesitated on account of the state of health which early self-abuse had entailed. He suffered from frequent involuntary emissions when asleep, and from the consequent loss of mental vigor and physical strength which usually follow. Alarmed on the account, he sought advice and medical treatment. He complained of a dull pain in the region of the lumbar vertebræ, and flaccidity of the penis. He suffered from tedious constipation and hemorrhoids, that often protruded from the anus ; *nux v.*, 3d trituration, was first prescribed— a three-grain powder every night half an hour before retiring—after which he was somewhat relieved of constipation ; but there was no improvement in other respects. *Sulphur* 6th, was then substituted for the *nux v.* and continued for a week ; no good result followed ; *conium mac.*, 6th dilution, ten drops in half a tumbler of water, was directed to be given in tablespoonful doses, morning and evening. He soon felt better, and after two weeks the hemorrhoids disappeared, the pain in the back was better, and normal virility returned. A short time after the young man got married, and has lived happily with his wife for the last five years, and is the father of two children, a son and a daughter.

12. *Conium*, 3d, 6th and 30th attenuation have been advantageously employed in the treatment of nocturnal emissions, or those that occur involuntarily at other times, when there is a flaccidity of the membrum virile and a sense of weakness and pain in the back.

CHAPTER VI.

SEXUAL EXCESSES AND OTHER CAUSES OF SPERMATORRHŒA.

By spermatorrhœa is understood the unconscious loss of the seminal fluid when at stool or when urinating, or at other times from the most trivial exciting cause, when there are no erections, but a mere flaccid state of the penis. The first cause of this weakening discharge,which we shall consider at some length, is SEXUAL EXCESSES, *both in unmarried and married life.* In the former a reckless roaming lust, and the frequenting of brothels, wherewith to become satiated by frequent indulgence, are the primary causes of genital weakness that result in the premature decline of manhood. The effect of excesses of this kind, for the sake of gratifying the mere lust for variety, is to bring upon the victim a train of evil consequences hardly to be enumerated. 1st, upon the vital condition of the general organism ; and 2d, upon those organs essential to the integrity of manhood. After a general debauch, there is a complete derangement of the functions, nutrition becomes impaired, and the entire body suffers ; and this is not all, the mind participates in the general wreck and the victim becomes polluted in soul and body throughout. Frequent repetition soon reduces the victim to the lowest point of physical and moral degradation. At first his digestion is impaired, and he resorts to a stimulating diet to encourage his lusts ; then follows constipation, hemorrhoids and other ailments that, in conjunction with a corrupt longing for sexual pleasures, so deteriorate and weaken the sexual system as to destroy all power of retaining the seminal fluid, and it passes off in spermatorrhœa as readily as feculent matter passes from the bowels in chronic diarrhœa. The state of his mind is even worse than that of his body ; sickened by his own indulgence he at last hates the opposite sex, and never conceives of it the idea of chastity. Such is the effect of commencing a career of indulgence of roaming lust, irrespective of the consequences. Is there no remedy for a reckless youth of this description till all is lost ? Before the habit is confirmed by repeated indulgence, it is possible to break off such a career and voluntarily refrain and reform, but as the habit becomes more and more confirmed by illicit intercourse with a variety of courtezans, there is the greatest danger of the complete wreck of manhood. Who is able to utter a successful warning before it is too late to save the victim from becoming a mere driveler and a show, with genital weaknesses that inevitably lead to absolute destruction of virility, and, a confirmed spermatorrhœa ?

2. The most inveterate of all maladies to cure are those found in such a wreck of humanity—the fruits of unconquered lust. Nevertheless, so long as a spark of humanity remains, or in other words, so long as any moral sense remains, a struggle to reform is possible. When a manly struggle is made to break off lewd practices, and the mind is withdrawn from lewd imaginings, there is at least some hope of recuperation and recovery from the effects of debauchery. If the digestion is impaired and the stomach irritable, and rejects the food taken into it, *pulsatilla*, 6th, in doses of four globules, three times a day, may be taken when there is a sense of weight in the stomach or a sense of contraction. *Nux vomica*, 6th, may be substituted for the *pulsatilla*.

3. After *pulsatilla* and *nux* have done their work, *china*, 6th, may be taken in the same way. In case of constipation and accumulation of hardened feculent matter in the rectum, *lycopodium*, 6th, may be taken morning and evening in connection with a diet of digestible meats and vegetables, with bran bread, fruit such as apples and pears, and no stimulating condiments. Great regularity in taking the meals, and

the repudiation of late suppers, oysters, etc., are essentially necessary.

4. In case of hemorrhoids that become inflamed and affect contiguous tissues, so as to produce or augment spermatorrhœa, æsculus globra tincture may be taken in drop doses in a spoonful of water, and repeated every three hours. In very many cases this remedy will remove the piles, and if the spermatorrhœa remains and the seminal fluid passes off when urinating or straining at stool, *selenium*, 3d, or *conium*, 3d, may be dropped in the proportion of ten drops to half a goblet of water, and a tablespoonful dose may be taken every four hours during the day. A strict observance of the above course will accomplish much in recovering the lost manhood.

5. For varices or tumid veins that become so sore and painful as indirectly to produce weakness, if not paralysis of the seminal vesicles, *arnica* and *pulsatilla* are remedies to be consulted.

6. Affections of the rectum and anus that have resulted from other causes than sexual excesses, may produce spermatorrhœa, and before the latter can be cured these affections must be removed, and well chosen remedies for the particular troubles often have a salutary effect.

7. Retained fæces, by pressing upon the seminal vesicles, induce a semi-paralytic state of the muscular coat, disabling them so that they cannot resist the pressure of the fæcal masses when at stool, and the seminal fluid passes off involuntarily. Similar effects may occur during the act of urinating by the contraction of the bladder. In every case, therefore, of spermatorrhœa, the state of the bowels and the condition of the urinary organs should be critically looked after, and such remedies must be selected as will be most likely to remove the proximate cause, and in a majority of cases *nux vomica* or *lycopodium* will suffice ; a dose of either every night may produce relief.

8. Fissures of the anus, which are accompanied by cramplike contractions of the sphincter ani, and pain at

stool, tend to retard the evacuations and cause retention of fæces. The itching and tickling of these are of great importance in explaining many cases of spermatorrhœa. The intolerable itching compels scratching the perineum and anal region ; the adjacent testes suffer from daily irritation provoked by this act. The organs contiguous participate in the irritation, and this is often followed by weakness and spermatorrhœa, without sexual excitement. *Sulphur*, the 3d trituration, taken in 3-grain doses, every night, will generally cure the fissures and remove the itching, and consequently effect a cure.

9. Herpes, which consists of numerous minute vesicles upon the scrotum, penis or perineum, and around the anus, may burn, itch and smart, and provoke the patient to rub and scratch until the genitals become so weakened that spermatorrhœa may be occasioned. To remove this condition, *petroleum* has been found a specific.

10. A gentleman of steady, temperate and virtuous habits, was almost maddened by this herpetic eruption, and spent much time and money in striving for relief, was at last advised to take *petroleum*, the 3d attenuation, five drops in a spoonful of water, night and morning, and in less than two weeks he was entirely cured.

11. When the seminal vesicles are no longer able to serve as reservoirs for the seminal fluid by reason of the constant irritation which produces depression and weakness, and on the slightest provocation discharge their contents, there results an habitual spermatorrhœa. *Petroleum* as above may be given first, and afterward *sulphur* to cure the difficulty.

CHAPTER VII.

SPERMATORRHŒA CAUSED BY ASCARIDES AND GRAVEL.

The effect of thread worms upon the sexual sphere is both dangerous as well as distressing. They generate in the large intestine, and particularly in the rectum, and to this circumstance must be attributed the impotence of some persons who, through their annoying influence, have been initiated into self-abuse, and thence into suffering from seminal losses and spermatorrhœa. Very young boys often become the victims of suffering from these parasites, and comparatively a less number of adults find themselves victimized by their presence.

2. This inconsiderable worm is cylindrical and pointed, white in color, about the size of common wrapping twine used for tying up small parcels. The tail end of the male ends abruptly, and is rolled up in a spiral, while in the female it is straight and pointed.

3. The head is provided with wing-like attachments, between which the mouth is situated; the length varies from a line to a line and a half. It sometimes appears in the evacuation in large numbers, so great even as to present the appearance of a constant wrigling motion.

4. When we consider the great number of symptoms this worm is capable of producing, such as tickling the nose, squinting, colic, fits, etc., it is not difficult to comprehend the nature of its action upon the anus in producing that intolerable itching which is so hard to endure, and not only the anus, but the intestines, testicles and penis, and the most sensitive seminal vesicles, exciting them to involuntary emissions by self-abuse, or to pollution and spermatorrhœa.

5. Thread worms produce nearly the same symptoms as stone in the bladder; children two or three years old suffer constant erections from them. This symptom affords one of the most positive indications of the presence of pin worms, and these boys grow up with the constant habit of handling the parts that itch, and scratching and rubbing habitually comes up with them in some form of self-abuse.

6. The opposite sex likewise suffer in the same way, and are led into habits of self-abuse, until health disappears and beauty fades. This constant itching and scratching irritates the skin, causes the clitoris to be reddened and swollen, and sometimes an ichorous discharge

from the vagina, and this only augments the itching, burning and swelling of the labia, all of which are powerful influences in augmenting self-abuse. We shall discuss this matter more fully when we enter upon the chapter relating to females entirely.

7. A learned authority, in treating of this particular affection and its consequences, gives as an indication of the presence of pin worms, deeply-sunken eyes, surrounded by blue rings. This appearance persists as an outstanding sign of self-pollution, even after the habit has been subdued and only its consequences remain ; and what are the consequences that remain after the habit has been corrected? We will see if we can divine.

8. Hypochondriasis, impotence, congestion of the brain, apoplectic fits, may all occur from the irritation produced by these apparently insignificant vermin, though all indulgence in self-abuse is done away with, and moreover all these affections have been cured when the worms have been removed. Whenever a boy or girl is found to be suffering from epileptic fits, inquire critically into their habits, and whether they have been sacrificed to the ascaris. Should it be ascertained that pin worms were preying upon them, correct self-abuse, destroy the worms, and nine times out of ten the epilepsy will prove to have been mere epileptiform convulsions caused by these iniquitous thread worms, and the patients will speedily recover.

9. A celebrated writer mentions a case of frequent nocturnal emissions of six years standing, which recovered rapidly after the removal of the pin worms, which it seems had been the cause. For all other means had previously failed, and the rapid convalescence of the patient after their destruction, proves them to have been the cause. This case, continues the writer, was also characterized by apoplectic symptoms and a partial loss of memory.

10. This writer is of the opinion that young men who are suffering from diurnal emissions, without erections or pleasurable sensations, were victims of enuresis dur-

ing their childhood. In such cases the disease consists in great irritability and weakness of the bladder, increased by the warmth of the bed, and conducted from the bladder 'to the neighboring sexual apparatus.

11. The neck of the bladder is the most sensitive when stones are trying to make their exit, and from this source the most intensely painful symptoms proceed. The more the disease we call "gravel" is prolonged, the longer the durations of· those affections of the bladder, and so much more extensive is the invasion of the contiguous parts or organs—the ureters, kidneys, prostate gland, urethra, rectum and vagina, all become implicated and somewhat exposed to danger.

12. The peculiar pains at the neck of the bladder, as well as at the base, while walking, sitting at stool, etc., and especially the pains experienced at the end of the urethra, induce the patient to violent pulling and stretching of the penis, and this leads to masturbation. Excessive length of penis, and an observable lengthening of the prepuce, as well as a thickening of the same, indicate the presence of stone in the bladder.

13. Sudden interruption of the flow of urine takes place when small stones are carried into the urethra and remain there. The reflex action of this irritation upon the rectum, vagina, testicles, kidneys, etc., as shown by cramp like contractions of the perineum, may produce an abnormal irritation of the whole urinary and genital apparatus. The tickling which stones in the bladder produce in the membrum virile often provoke involuntary emissions, and by the irritation being transferred to the seminal vesicles, testicles and seminal ducts, frequent pollutions and spermatorrhœa results, and so far as the physical health is concerned these may as well be brought on by masturbation.

14. A single large stone in the bladder, firmly imbedded in the fundus, may by its weight alone exert so strong a pressure upon the seminal vesicles that they must empty their contents and gradually suffer a diminution from atrophy, to the detriment and destruction of the generative function. In this way no other cause

than stone in the bladder may be the cause of spermatorrhœa and impotence.

15. The treatment of spermatorrhœa from these
causes consists in removing them. For that caused by
thread worms, such remedies must be employed as will
exterminate them. *Sulphur* is an important remedy to
give first—a dose of the 3d trituration may be given
every day for a week. This remedy may be followed
by *santonin,* 2d trituration, in the same way. Should
these not be sufficient, *terebinth,* in drop doses may be
given twice a day, and other remedies, *calc. cina* and
ignatia. A timely use of these remedies will in a majority of cases effect a cure.

16. To cure the spermatorrhœa that has been provoked by stones in the bladder, requires great care.
The cause must be removed, or otherwise there is no
chance for a cure ; when the stones are so large that
they cannot pass through the urethra while urinating,
and so hard and compact as to preclude the possibility
of their reduction by any other process, the sooner
some skillful surgeon performs the operation of crush-
ing them the better. It is preferable to undergo the
pain of lithotomy than to struggle a long time in such
suffering. After the operation has been successfully
performed, the after treatment with remedies must be
in accordance with the symptoms. The soreness and
pain consequent upon the operation is very soon relieved by *arnica,* 6th, ten drops in half a tumbler of
water, and a tablespoonful may be taken three times a
day. Urging to urinate or painful urination requires
cantharis to be prepared and taken the same as *arnica.*
Urethral inflammation calls for *cannabis sat.* It may
be that these remedies will cure the seminal weakness
after the cause has been removed. But if they fail, it
must be attributed to the extensive injury which the
seminal vesicles have already received. The persistent
use of *china* may in time bring up their vitality.

17. A young gentleman who had always lived a correct life, was somewhat rheumatic. He observed that
his urine deposited a reddish sediment, which stuck

fast and adhered to the vessel. In process of time he passed what he termed red sand when he urinated. He finally began to suffer from strangury and irritation of the urethra. At last he felt stinging when he urinated that extended to the end of the penis. Now, in our opinion, if this young man had taken at that time a few doses of *tart. emetic*, he might have had less suffering in the future. But he neglected himself and the difficulty grew worse, until he began to pass blood with his urine, and the heat and irritation was communicated to the testes and seminal vesicles exciting pollution and spermatorrhœa. The concretions in the bladder became so massive that lithotomy was a thing indispensable, and a skillful surgeon placed him in position to operate, put him under the influence of ether, and effectually crushed the deposits, so that they readily passed off with the urine, after which he took *arnica* and *cannabis* and felt quite relieved, but the pollutions remained and frequently the semen would pass off with the urine, and when at stool. *Digitalis* given in drop doses of the 3d dilution had a good effect. *Hypophosphite of lime* was given three times a day for a week or ten days, and the patient had no trouble afterward.

18. The diet in these cases should be barley water for drink, fruits, esculents, and good digestible meat and fish; all food tending to bind up the bowels so as to render constipation habitual, must be avoided.

CHAPTER VIII.

EXCESSIVE SEXUAL INTERCOURSE IN MARRIED LIFE.

Marriage is intended for a higher, holier and happier purpose than unlimited indulgence in sexual intercourse. It is not a license for excessive coitus, and therefore when made such the worst of consequences may follow. 1st, weakness of the general system; 2d, weakness and derangement of the function of reproduction.

1. Weakness of the general system is produced by

a series of disturbances which we will proceed to explain. When a man and wife cohabit for the sake of the legitimate purpose of begetting offspring, there is but little tax upon the strength or but little risk of impairing the generative function. But when the mind sinks into the low plane that craves sensual sexual indulgence, it morbidly and selfishly anticipates the most unlimited sway for the passions, and to gratify them is liable to become the chief motive and delight. Now there is a normal delight in sexual intercourse altogether different from the morbid craving for indulgence. The former is chaste and pure when it spontaneously occurs from mutual love and affection between married partners, and virility is constantly strengthened for the purpose ; with such the love looks to the important result—the legitimate fruits of marriage, and rejoices when children are born as the mutual pledge of connubial affection. But the latter is a mere sensual lust that looks no higher than for opportunities to gratify it. It despises the idea of rearing a family, and the children that chance to be born as the consequence of this indulgence are not welcomed as blessings, but as necessary evils.

2. There is a limit to this morbid and selfish lust. The gratification of it does not strengthen connubial tenderness and affection, nor promote the health of the parties. After a time the love grows cold, or is turned into hatred, and frequent and continual indulgence ends in the deterioration of mind and body.

3. The man from continual losses of semen finds his digestion impaired, his nervous system weak, and what is worse he finds himself the victim of sexual weakness and his virility impaired. His wife at the same time has become hysterical and fretful, and there is no happiness in the household. The man and the woman only come together when lust excites them to an embrace. At other times their backs are turned to each other in disgust and hatred.

4. Continual cohabitation at length destroys the function of the seminal vesicles, and paralyzes the little

muscles that prevent the escape of the seminal fluid, and thus manhood becomes sterile, prostrated and the victim of spermatorrhœa, and the woman has become his sterile companion. Excessive sexual indulgence has been the proximate cause. When the digestion has become impaired, and the successive chain of organic functions participate in the misfortune, nutrition becomes feeble, and the whole body suffers from emaciation and debility. The effect upon the woman is quite similar. She suffers from nausea, debility, and general nervous prostration. The picture is not overwrought. Excessive sexual indulgence, even in married life, results in the premature decline of manhood. It pollutes the soul and fills the mind with diverse fancies—it destroys the vital elasticity of the muscles, saps the nervous system and entails many weaknesses, such as rheumatism, constipation, hemorrhoids and renal disturbances.

5. When all the vital functions become thus impaired from over-indulgence in sexual intercourse, the query arises, Is there no remedy? Is the restoration of strength possible, and can manhood be restored? If not too low or too far gone we answer these questions affirmatively.

6. When one sensibly feels that his virility is waning, let him pause and consider, let him direct his mind and thoughts to a higher plane of love and affection, let him refrain from indulgence and lust and turn his back upon the wicked practice. If he feels feverish and restless, let him take a few doses of *aconite*. If his appetite is impaired and his food distresses him, let him take *nux vomica*. If his back is weak, rheumatic and stiff, let him take *cocculus*. If his bladder and urethra are irritated, let him take *cannabis;* or if he has strangury, let him take *cantharis;* and for general weakness, flaccidity of the penis and spermatorrhœa, let him take *china* persistently, and eat and drink—if his appetite permits and nutrition is not completely interrupted—well cooked meats and vegetables and drink wholesome drinks. If he fulfills these conditions with-

out relapsing into more selfish indulgence of his passions, virility may be restored.

7. The worn out and depressed wife must also direct her mind in that channel most conducive to her happiness. Let her thoughts and affections ascend and rest in a religious view of married life. If she suffers from nausea and indigestion, she may improve the condition of her stomach by taking 3 grs. of the 3d trituration of *oxalate of cerium.* This remedy will strengthen her nerves, improve the digestion, allay the nausea, and give general tone to body and mind.

8. When both parties have thus complied with the means of regaining health and strength, they will be able to come together as man and wife, and with lofty sentiments above venery they will happily find connubial love and affection to take the place of lust, and they may come into the happy relationship of husband and wife. If otherwise, they will sink lower and lower, the victims of excessive lustful indulgence. Connubial bliss and conjugal tenderness will bloom no more for the household.

CHAPTER IX.

THE CONSEQUENCES OF ABNORMAL SEMINAL EMISSIONS.

Seminal emissions are either abnormal on account of the means by which they are brought about, or in respect to the frequency of their occurrence, whether produced by coitus or spontaneous pollutions at short intervals.

2. Long-continued and oft-repeated masturbation in both sexes is altogether abnormal, and is the fruitful source of disease. In the male it results in disease of the reproductive organs, and is followed by emaciation and consequent debility of the whole body. Hippocrates maintained that this emaciation indicated the atrophy of the spinal cord. He describes the sense of formication, or the feeling, as if ants were crawling over the skin. as an accompaniment of the atrophy,

and the loss of seminal fluid while urinating and when at stool as the result, and to this is added a sense of weariness and shortness of breath after a short walk. All this may occur when a person of the most robust health is broken down by the vice.

3. Celsus supports the views of Hippocrates and maintains that atrophy of the spinal cord is the immediate source of emaciation and the legitimate consequence of abnormal emissions, produced by masturbation. It has been observed that the emaciation of onanists increases in spite of a good appetite and the consumption of a large amount of food. Insatiable hunger and a good digestion are symptoms that indicate the struggle of nature to compensate for the losses, and yet so long as the spinal centre is the source and its atrophy stands out in continual decrease of the flesh, satiated hunger and an unimpaired digestion can avail but little. This emaciation is often ascribed to the rapid growth of youth that sometimes follows puberty. The muscles of the hip and lower extremities show forth this peculiarity, which is ascribed to pathological changes of the spinal cord.

4. Just in proportion to the emaciation the onanist loses his strength. He leaves his bed with difficulty in the morning, and is dull and listless during the day, even when at his work. In going up stairs he finds difficulty in breathing and palpitation of the heart. These symptoms of weakness may increase to an alarming extent, till the onanist bends over like an old man and faints and reels from vertigo on the slightest exertion and is obliged to keep his bed.

4. The case of the onanist, even in the extremity described above, is not hopeless. The strength and fullness of the body may return when the vice is given up and proper remedial means are employed to obviate the deterioration of the spinal cord. The habit broken, the appetite and digestion good, render it probable that the 3d decimal trituration of the *hypophosphite of lime*, given persistently for a sufficient length of time, will restore the spinal cord to its nor-

mal size and strength. The remedy may be administered in 3 grain doses, half an *hour after* each meal
and before retiring.

5. Modern physicians concur with Hippocrates and
Celsus in these descriptions of the consequences of
seminal losses by onanism. Hoffman says: "The
onanist loses his strength after frequent seminal evacuations, the body gets thin, the face pale, the memory blunted or lost, and a continual coldness seizes
the limbs ; the face becomes idiotic, the voice hoarse,
and in short the whole body is reduced to atrophy,
and sleeplessness, restlessness and tormenting dreams
are the usual concomitant symptoms." The same author says: "Amaurosis or total blindness is sometimes the consequence of abnormal seminal evacuations."

6. Boerhaave has noticed "pain in the membranes
of the brain, weakness of the body, blunting of the
senses, leanness and paralysis, and this is not all. The
face loses its healthy and beautiful tint, becomes pale,
earthlike and yellowish, or lead colored and livid ; the
lips pale and the eye dim and glassy. The bluish
margin around the eyes, a puffiness of the lids, flabbi-
ness of the flesh, weak and small pulse, copious sweats,
swelling of the upper and lower extremities, and finally
hectic fever and general symptoms of exhaustion are the
deplorable consequences, which show that the organism does not succumb to the onslaught on its integrity
without the most obstinate struggle." To change this
condition requires, as before stated, a complete cessation of the polluting habit and a steadfast resolution
to rely on the best regimen and remedial means to restore the body to its normal health.

7. At first the abuse of the genital organs begets in
the onanist a sense of hunger and a voracious appetite,
but long-continued sexual abuse results in indigestion,
loss of appetite and disgust for food, or at best the appetite becomes irregular, vague and beset with morbid
cravings and derangement of the sense of taste. Food
taken into the stomach causes pain and vomiting, or

diarrhœa and flatulence or constipation and hemor-rhoids. For this condition *sulphur* and *nux vomica* may be given as follows: Upon the supposition that the victim is alarmed at the consequences and has broken off the habit which induces abnormal seminal losses, give *sulphur*, 3d, every night for a week, and then follow with a dose of the 3d of *nux v.* every night just before retiring, and with such a diet as will accord with the condition of the stomach, the above conse-quences in many cases may be obviated.

8. Abnormal seminal emissions are said by Deslandes to lead to other diseases, classed among the severer and fatal forms, such as apoplexy, ramollissement, epi-lepsy, chorea, mental disorders, spinal irritation, blind-ness, deafness, gout, strabismus, varicocele, sarcocele and hydrocele, many of which are incurable, and they must therefore remain as a permanent warning to young men not to make shipwreck of themselves upon the rock upon which so many have foundered and sunk.

9. Another consequence of onanism is the complete decline of virility and inability to propagate their kind either because they are unable to perform the marital act or because they have lost all the warmth which healthy semen requires to vivify the female germs. Should such enervated individuals beget children, upon the principle that "like begets like," they will be feeble and puny, and as they grow to maturity they will be ill-shaped, bow-legged, oldish-looking specimens of humanity, and victims for an early grave. The more healthy, strong and sound the father, the more robust and perfect will be his offspring.

10. Bodily diseases alone are quite enough to utter a warning to the onanist. But they are trifling when compared with the awful consequences upon the soul. The mind succumbs, childishness and imbecility are its attributes, and he becomes a moral monstrosity; thank heaven, monsters cannot propagate, neither physical or moral, for manhood is gone, virility destroyed, and the vessel is a complete wreck in the sea of human infirmi-

ties. Such are the consequences of this kind of abnormal seminal losses.

11. As stated in the preceding chapter, a high, licentious degree of sexual abuse results in hopeless degeneracy of the organs of reproduction, as well as of the whole animal system. Their excitability gradually diminishes, and mental disorders, associated with their diseased condition, become prominent. This is especially the case when the body is well nourished with good food, while at the same time the debilitating cause continues to act on the genital organs. Pangs of conscience, remorse, shame or fear of the terrible consequences, as set forth in certain books, easily excite apprehension, melancholy and hypochondria. When the mind constantly fights against the disease-producing lust, it is in a continual state of excitement. The brain becomes affected, and mental disturbance or insanity is superinduced upon the physical weakness, and in the lust it changes to hopeless idiocy.

12. The conviction of incurable impotence, joined to intemperate habits resorted to for the purpose of silencing anxiety and fearful forebodings, frequently contributes to unsettle the mind and pave the way to hopeless dementia. One-tenth of all the inmates of our insane asylums are of this class of secret sinners, who still continue the vile habit of self-abuse, though brought to spiritual and moral bankruptcy. Insanity, hallucinations, loss of sight and hearing in a moral point of view form the climax, characteristic of the consequences, and withal these idiotic victims of sexual abuse embrace every secret opportunity to instinctively cultivate and practice the vile habit of onanism.

13. The consequences of masturbation in the female sex are equally disastrous and injurious to the general organism. Some eminent physicians maintain that the delicate and susceptible organization of the female system renders it more liable to suffer from this vice than that of the male. Suffice it to say the female masturbater suffers all the physical and mental deterioration that the practice induces in the opposite sex,

and besides she becomes the victim of uterine affections of a serious nature, such as disturbance of the menses, prolapsus, displacement, ulcers, indurations, and cancer.

14. One of the most serious consequences of the habit is local irritation of the nerves of the womb, resulting in nymphomania, which affects both soul and body, and degrades the finest feelings and attributes of her being—disgusting to herself, and a shock upon female or womanly modesty.

15. Rozier, a French physician, asserts that masturbation in girls, by the frequent and powerful cramplike contractions accompanying the fulfillment of the act, induces considerable swelling of the neck, as in epileptics ; and further, that in some the skin becomes yellow, and in others eruptions resembling ringworms make their appearance on the arms and legs, which disappear when they refrain from the vice, but return when a repetition is indulged in. The voice, also, of such girls becomes rough and hoarse, hollow and weak, losing its sonorous, soft and metallic ring.

16. A feeling of oppression in the chest and region of the stomach, with a dragging, occasionally indicating the need of food, are consequent on the habit. Severe cramps in the stomach, and similar disturbances of the solar plexus especially, show themselves in girls who indulge in this habit, and at the same time leucorrhœa becomes established, attended with other troubles, such as spasms and cramps in various parts of the body, eructations after eating, distention of the abdomen, difficult digestion, headache and restless moving of the limbs.

17. A sallow countenance and an ugly expression of the face, which is pale and sickly, are signs which betoken the physical and mental depression, consequent upon masturbation. The vivacious expression of youth gives way, the eye becomes dim and surrounded by leaden colored rings, the lips are pale, the teeth covered with a gummy and dirty-looking mucus. The entire fullness of feminine spirit and beauty has seem-

ingly vanished, and the shrunken image betokens premature decline. Both heart and mind suffer more in comparison from the habit than is the case with the opposite sex. Worse than all are the pangs of conscience, and mortification, and grief which such a vile habit engenders. This, added to physical exhaustion, becomes prophetic of the shipwreck of all the glorious attributes of womanhood. The yoke under which she habitually labors is of such a nature, that the mind moves in a perpetual circle, and from which it cannot elevate itself. The whole endeavor is to mislead the eyes of others away from her true condition, and to recall the memories and imaginations which give fresh encouragement to the lascivious practice.

18. And yet another kind of moral deformity may spring up from the practice. The mind accustomed only to its own selfish broodings and lascivious thoughts, does not feel at ease in other spheres of thought. The pleasure of self-abuse then becomes the chief delight of those erring misses, and they give themselves up to it. All pleasure of concourse of the sexes is lost, and nothing but loathing and indifference takes its place. Their own silent, or rather secluded indulgence, eclipses the higher and nobler enjoyments. The sexual impulse with such is perverted and usurps dominion, and is much more frequent, according to Tissot, in women than in men. He cites the case of a wife who had become so confirmed in the habit of masturbation that she esteemed the pleasure superior to marital intercourse, for which she felt an unmitigated disgust. Tissot also remarks that this abominable habit keeps some girls from marrying at an age when they could do so; because, in their estimation, it would deprive them of this unnatural method of gratifying their passions, and hence the increased number of old maids.

19. It is true that the fluids lost by women by masturbation are less vital and perfect than that lost by men, and for this reason women can endure these exercises longer and more frequently without apparent injury. But the longer and oftener the woman gives

herself up to the practice, the more serious the consequences become, and this is attributable to the delicate
organization of the female system.

20. Among the examples of broken down constitutions, occasioned by the vice, we find recorded numerous instances of confirmed melancholy and insanity,
nymphomania, idiocy and suicide. Dr. V..Graifa, of
Berlin, relates a remarkable case of recovery from idiocy
in a young woman after the amputation of the clitoris
and the cessation of masturbation.

21. The TREATMENT required for impaired female
constitutions, so long as reason and moral sense remains, must primarily be moral—onanism must be dis-
carded, denounced and condemned. The mind must
be directed to higher aspirations and purer thoughts,
and for the debility and loss of strength, a good wholesome diet, and exercise in the open air are commended,
with two doses of *china* daily until relieved. For the
cramps of the incipient stage give *nux vomica* and
sepia 3d decimal, whenever they occur. For the weary,
tired and listless feeling, *arnica* third decimal, may be
taken twice a day. For nervousness and hysteria or
for timidity and spasm, *hyoscyamus* may be given three
or four times during the day. If dejected and inclined
to weep, *ignatia*. For epileptiform troubles, *cuprum
met.* Loss of mind and memory, *sulphur.* For nymphomania, *cantharis*. All of these remedies may be
employed in the 3d decimal attenuation, and prepared
for administration in the usual way.

22. Certain aliments are prohibited, such as oysters,
eggs, and indigestible meats, and all stimulating bever-
ages. By following the above directions all will be
accomplished that can be in the way of restoring sound
health.

CHAPTER X.

CONSEQUENCES OF SPERMATORRHŒA AFFECTING THE
WHOLE SYSTEM.

1. Spermatorrhœa, as before stated, consists of un-

conscious seminal emissions, that occur from the most
trivial excitement, distinct from the disease-producing
causes on which it depends. It gives rise to a series of
symptoms of great importance to the entire organic
structure of man. It is not only evident that the
causes of spermatorrhœa sometimes remain, but that
they affect pathological changes in different parts of the
genital organs. The disturbances which are wholly
due to the seminal losses are as follows :

2. The entire organism becomes altered and impaired ; the sufferer without being able to fix the locality of the pain, and probably without realizing the
nature of his trouble, is beset with general discomfort,
lassitude and trembling weakness of the extremities—
a depressed condition of the entire body, and a distaste
for any occupation of body or mind; not fully realizing
the nature of his trouble, nor able to explain, he retires
exhausted, sleeps indifferently, and awakes without
having been refreshed or improved in strength, with a
sense of pressure, fullness and dizziness of the head,
and inclination to fainting.

3. An enfeebled and sick body is but a feeble in-
strument for the mind that depends upon the cerebral
center for its integrity. It must therefore suffer when
the body is thus depressed. Its activity is seriously
impaired and morbid. The fire of the intellect can
glow but faintly when the whole physical system is in
such a flickering condition. For matter and spirit are
so closely and intimately related in human beings,
there must be a reciprocal influence of each upon the
other, or a perpetual conflict between nature and
spirit, body and soul, or matter and force.

4. And to add to the misery and wretchedness, a
knowledge of its source, or a self-consciousness of guilt
and the probability of having become the victim of
incurable disease, only opens the channel for fearful
and tormenting forebodings. Impotence is mortify-
ing, and to be in this hopeless condition fills the mind
with despair and leads to confirmed melancholy.

5. Then, further, a continual depression of spirits.

and brooding over this unpleasant condition, leads to intemperate stimulation, and this to affections of the brain, and a train of evil consequences that betokens a complete wreck of body and mind, the fruits of which may culminate in despair and suicide, or in insanity or idiocy, and the absolute loss of mental impressibility, and finally a sinking away in exhaustion, or an apoplectic convulsion, may end the train of evil consequences of self-abuse and spermatorrhœa.

6. There is another class of symptoms that sometimes shows itself as the consequence of abnormal seminal losses. This class embraces great muscular weakness, and different forms of paralysis. There is an intimate relation and mutual dependence that exists between the nervous system and the blood; nervous energy depends on the purity and normal condition of the blood. An impoverished condition of the circulation explains the diminution of muscular power. For it is self-evident that the *vis nervosa*, which is the natural stimulant of muscular power, must cease to be active, in the degree that it fails of support from the blood. Spermatorrhœa not only deprives the circulation of the purest elements of the blood, and thus produces an irritable condition of the nervous system, but by reflex irritability the nearest spinal nerves and thence the whole cerebro-spinal system may become affected; and by continuity the muscles become subject to extreme weakness and paralysis. As for example the lower extremities sometimes exhibit a semi-paralytic condition, causing the victim at every step to throw his legs, irrespective of muscular control.

7. Paralysis of the bladder, rectum and anus follow, and the muscles of the hip also become implicated. Sometimes only a single locality becomes paralyzed, and this may be the tongue or perchance the sphincter of the bladder or anus, and this depends upon the extent of the irritation of the nervous system.

8. When brain affections are becoming more extensive, one organ after another may show the effect in partial paralysis. When stuttering and stammering or loss

of speech entirely manifest themselves, we are led to suspect some affection of the nerves immediately connected with the brain, on which the muscular apparatus depends ; and it does not necessarily follow that other localities will become similarly affected.

9. But inasmuch as spermatorrhœa being in some instances, the primary cause of paralysis of the tongue, this symptom has been regarded as ominous of more serious troubles, when the losses of semen are persistent. It may be the first alarm of an increasing or progressive paralysis that may terminate in general mania, and on this account the symptoms have been studied in lunatic asylums as the prodromous or forerunner of alarming results.

10. That the most frightful cases of chorea arise from self-abuse and seminal losses, no reasonable doubt obtains. The irritation of the brain which results therefrom, causes a semi-paralytic condition of the facial muscles, which gives numerous twitchings and a peculiar expression to the face, and if not obviated there may arise a more extensive paralysis implicating the optic and auricular nerves, and those of the palpebral muscles, rendering it difficult to open or close the eyes.

11. But the effect does not end with mere weakness of sight. Rognita describes a case of amaurosis from excessive seminal losses in a young Jesuit from Palermo, who indulged in self-abuse six or seven times a day. Deslandes considers the amaurosis a symptom of great exhaustion, and of a parallel character with that of the legs from spinal irritation. In addition to this blindness, the motor muscles of the eye may become sadly affected, and strabismus and spasms of the greatest intensity may take place, and also a constant lachrymation and agglutination of the eyelids in the morning.

12. It must be confessed that spinal irritation is one of the most disagreeable consequences of spermatorrhœa. It is a diseased condition intermediate between nerve pain and inflammation, and is denoted by a press-

ing, drawing sensation in the region of the hips and small of the back ; disturbed sensibility, formication, alternation of coldness and heat, pressure and weight, and pain in bending down, and through the enfeebled genitals the irritation extends to the lower portion of the spinal cord, and is reflected to the testes and penis, producing in them a sensation of drawing, pressure and dullness, extending to the inguinal and hypogastric regions, and from thence it may be reflected upwards to other organs, producing an uncontrollable restlessness.

13. A super sensitiveness of the entire organism is liable to result. The auricular nerves are too sensitive to endure loud talking, music and the like. The subject is unable to concentrate his thoughts on any subject. He can neither endure the act of reading or writing, and he must therefore remain inactive, beset with sleeplessness, headaches, dizziness, perverted taste and smell, and pricking and itching of the skin.

CHAPTER XI.

THE EFFECT OF SPERMATORRHŒA UPON THE RESPIRA-
TORY SYSTEM AND THE HEART AND OTHER ORGANS.

1. In the course of the general irritation of the nervous system, arising from spermatorrhœa, the breathing apparatus becomes implicated, and oppression of the chest and præcordial anxiety weigh heavily upon the patient. A short, dry, and persistent cough most always results from the sensitive condition of the pulmonary and bronchial nerves, and palpitation of the heart sets in to complete the picture. Nearly every authority upon the causes of sexual diseases allude definitely to the asthmatic symptoms produced by pollutions and spermatorrhœa ; associated with these are dry cough, debility, feebleness, restlessness, perspiration and stitches, which actually imitate the symptoms that usually accompany tuberculosis. Many a practitioner has been puzzled to institute a diagnosis, to tell the difference.

2. Diseases of the heart and large blood vessels frequently result from self-abuse and spermatorrhœa, and sometimes from excessive sexual indulgence. Nevertheless, of organic trouble of the heart may be present without any lesion. In such cases the appearance is due to spinal irritation alone, which extends in a greater or less degree to the medulla oblongata and cerebellum, and this also explains the unbearable pain and sensation of pressure at the nape of the neck and back of the head, and the tendency to bend backwards as in opisthotonos of the neck in transient tetanus.

3. Gouty pains in the feet and knees, or hands and fingers, in conjunction with convulsive movements and trembling are also consequent upon self-pollution and spermatorrhœa. Epilepsy and chorea frequently follow sexual excesses—the latter more frequently in young girls troubled with thread worms, who, through their biting and itching influences, have led to rubbing and masturbation.

4. Through disturbed innervation the nerves of the stomach become implicated, producing pains, cramps and spasms, and through the general deterioration of the body indigestion and defective nutrition may result. The torpor of the muscular system throughout all the organs retards or depresses the vital activity of all the organic functions, and loss of appetite, dyspepsia and depression of spirits are certain to render mind and body uncomfortable.

5. The vitality of the skin becomes impaired and its function interrupted, and when distended by foul gases in the stomach, and wind, the condition fully accounts for the severe colic and cessation of peristaltic action, accompanied by constipation from which the victim continually suffers. Constipation is one consequence of spermatorrhœa, and it serves at the same time to stimulate a renewal of seminal losses.

6. Constipation and hemorrhoids, which frequently exist at the same time, are probably classed among the products of continual seminal losses, and also other affections implicating the neck of the bladder and prostate

sufficient to oppose the ejaculatory forces, and prevent the flow of semen during the act of copulation; such obstacles are difficult to overcome.

13. Atrophy, and cancerous induration of the testes are most unfavorable in regard to prognosis. Sarcocele, varicocele, and hydrocele inevitably lead to impotence, because the secretion of the seminal fluid in such cases is hindered, or sexual intercourse is both painful and difficult. Spermatocele, which is a swollen condition of the scrotum, resulting from an accumulation of seminal fluid in the testes, epididymis and vas deferens, either through voluntary retention of semen in copulation or through abstinence, is undoubtedly followed by impotence.

14. Such are in general and in particular the causes of impotence, and nearly all of which may be traceable directly or indirectly to some form of self-abuse, and such are the consequences of pollutions and spermatorrhœa, as seen in the premature decline of manhood. The picture drawn in the foregoing of the disastrous consequences of pollutions and spermatorrhœa, although correct to a certain extent, must be viewed more as a warning to those addicted to self-abuse, than a source of discouragement to those unfortunately afflicted, and therefore in the following chapter we shall treat of marriage in relation to sexual weakness.

CHAPTER XII.

MARRIAGE, IN RELATION TO SEXUAL WEAKNESS.

1. When a morbid imagination has led to sexual abuse, and the whole sexual system has become impaired thereby, the victim is no sooner aroused to a sense of his condition, than a morbid and discouraging fancy begins to influence him in another direction, and he too frequently regards himself the victim of incurable disease. But this is not warranted.

2. When a young man who from some cause or influence had been initiated into solitary habits of self-

abuse begins to think seriously upon the consequences, he is apt to imagine himself unfit to assume the relation of husband to a wife, and under a sense of remorse he broods over his situation until he dreads the future, and hesitates when he looks upon marriage as desirable. But there is in the main no occasion for this, and the sooner his will can triumph over these morbid forebodings the better.

3. When he comes into a state to renounce and denounce as wicked and disorderly the habit of self-pollution, he takes the first step to regain his manhood. If he suffers from sensible weakness on account of what has happened, it behooves him to employ the best remedial measures, with hopeful reliance on them for a cure. Looking forward to matrimony is as likely to benefit him as any means he can employ, provided his motives for entering into such a state are right, and he desires to become an affectionate and faithful husband. Even if some of the effects of his former indiscretion remain, the marriage relation is as likely to favor his entire recovery from them, and even more so than if he remains single.

4. In a happy married life the incitement to sexual intercourse being normal and springing from affection, has a tendency to strengthen mind and body for the purpose. The affection of such a man for his wife, who fully requites his love, has an undoubted tendency to strengthen the sexual system. We have known instances of seminal weakness so great as to excite apprehension and alarm, to entirely pass away after marriage. But in such cases much has depended upon the previous exertion of the will to fix the mind upon chaste subjects and to avoid all excesses and broodings over the past. A young student of the university, subject to nocturnal emissions four or five times a week, found himself in a failing condition of health, and without ability to concentrate his mind upon his studies. He applied for advice, and medical treatment. He became very despondent and imagined for himself the worst of future consequences. To encourage his hopes, and

direct his mind to chaste subjects, he was advised to turn his attention to the subject of marriage, and to look forward to such an event for himself; to which he replied that his indiscreet habits had ruined him and blasted his hopes in this direction. Although he had fully broken himself of masturbation, the evils entailed was what beset him, and interfered with his health and peace of mind, and he had therefore concluded to abandon his studies, and try to recover himself in some secluded way. Remedies were given him, accompanied by encouragement to rest awhile, and try the " Health Lift." He did so, and derived great benefit. After a season of rest he returned with ability to complete his studies, not cured of his infirmity, but greatly improved; after which he left the university and went into business, and soon became engaged to a lovely lady for whom he cherished the purest affection. But he hesitated and delayed entering upon marriage until advised that such an event might obviate and cure his emissions, and be the means of restoring rather than of diminishing his sexual ability. He finally took courage and entered into wedlock and became a happy husband, and in due time the father of several children, and after was never troubled with the weakness that had so preyed upon his mind.

5. The above is by no means a solitary example. Marriage from pure motives, and not as a means of gratifying lust, is ordained of heaven to be the means of strengthening and perfecting the powers of manhood, and getting rid of many evils incident to a bachelor's life. Love is not lust, and in a beautiful and affectionate wife it brings love in return, and this love is life, and full of power to overcome weakness and give legitimate strength to virility.

6. But to look forward to marriage as a license to whoredom must be corrupting to the wife, and a source of greater weakness and suffering to both parties. In this instance lust takes the place of love; and as lust has previously led to excesses and self-abuse, to marry for the sake of affording it unlimited indulgence is only

adding fresh fuel to the fire, and the physical and mental strength diminishes, and peace, love and affection depart from the household.

7. We will therefore say to all young men that marriage from pure motives is honorable, and though the follies of youth may have preyed upon your health and brought on pollutions and even spermatorrhœa, you are not lost—your manhood is not gone—provided you exercise the power of will to break off all lewd habits which a morbid imagination begets, and turn your attention to true love and marriage; for love, requited and pure, is the fulfilling of the law of marriage; it can never lead to sexual excesses, but by the employment of judicious measures in connection therewith, it may give fresh life to the mental and physical powers, obviate disease, and cure seminal weaknesses. In order there is beauty and strength—in disorder there is confusion and weakness.

CHAPTER XIII.

RECAPITULATION AND TREATMENT OF SEXUAL WEAKNESSES.

1. In the foregoing chapters we have enumerated the causes that operate to produce the premature decline of manhood, and the numerous effects that proceed from these causes. We have also given some general therapeutic hints concerning remedies. In this chapter we shall conclude the work by a brief recapitulation and special treatment with diet, regimen and remedies.

2. In all cases the cause, whatever it may be, must be removed if possible before the effect can cease, and causes are of two kinds, viz.: primary and secondary.

PRIMARY CAUSES are those which primarily act upon the general health, inducing functional or organic derangement.

SECONDARY CAUSES are the conditions that immediately influence, aggravate or induce diseases of the seminal vessels.

3. Among the primary causes of masturbation with the young of both sexes, we have seen that worms and eruptive difficulties, that occasion much itching and consequent rubbing and scratching, are to be included; and it is incumbent on parents to be exceedingly particular with their children at this tender age, in order to guard them against such initiative influences.

TREATMENT.—For thread worms, *sulphur, santonin* and *terebinth.* have each proved successful in removing them. The *sulphur* may be given in the 6th dilution, a dose every twenty-four hours. Should this fail, follow with *santonin* 3d, morning and evening, or with *terebinth.* 3d, morning, noon and night. Give the *sulphur* and *terebinth.* in drop doses in a spoonful of water, and the *santonin* in powder. This treatment will often suffice to arrest the effects of those annoying parasites. To cure the eruptive difficulty and relieve the itching, *petroleum* 6th in drop doses three times a day will be found useful, or else *calcarea conium* and *sulphur.*

When girls at a tender age have been initiated into masturbation by such annoyances, serious consequences have arisen. We have recently seen a sad case of chorea which resulted from these insignificant parasites, first initiating the habit of rubbing and then of masturbation. She was cured of the malady by *terebinth.* 6th. She was past nine years of age, and after becoming relieved of the thread worms, her general health and strength was greatly improved. Other cases have been cured by *sulphur* given persistently every night for a month. *Santonin* after *sulphur* will generally exert a healthy influence upon the mucous membrane, and entirely obviates the itching, and therefore one of the primary causes that initiates into the habit of self-abuse becomes removed. Itching from some eruptive disease on the integuments of the genital organs, is another primary cause of self-abuse and sexual weakness. This eruption has been cured by *petroleum*, and the itching entirely subdued. *Conium mac.* has been successfully employed for the same purpose, and so has *sulphur*. The two former, when required, may be given in the 3d dilution three times a day, and the latter when required may be given in the *tincture* every twenty-four hours.

WHEN SEXUAL WEAKNESS is primarily caused by masturbation, there is little hope of cure until the mind, the thoughts, motives and sentiments become set against the habit, nor until the mind becomes elevated above that condition which a morbid imagination engenders. To come into this state requires a strong will, and in youth, the kindest encouragement from friends.

If, as a result of self-abuse, there occur nocturnal emissions, causing a sense of debility and dullness, *china* 3d dilution may be administered three times a day before each meal, or until the sense of debility is removed, or *plantago major* 3d dilution may be given in the same way. Where there is a feeling of malaise and confusion after excessive emissions *phosphoric acid* dissolved in water, the 3d decimal in 5 drop doses may be given morning and evening until the malaise and confusion are better. *Selenium* 3 may be given instead if there is vertigo on rising in the morning, or there has been an oozing out of semen when asleep, or a discharge of prostatic fluid. *Sepia* 6th to the 3d will cure excessive nocturnal emissions when they are followed by hypochondria, weak memory, sadness, depression of spirits, dullness of the head, and weakness of the sexual organs. When constipation aggravates the discharges, or excites them *nux vom.* 3d or 6th may be given to overcome the difficulty. Dr. W. H. Burt cured several cases of spermatorrhœa, attended with much nervous irritability, with half grain doses of *bromide of potassium*, repeated every six hours for several days. *Canuabis sativa* has been prescribed successfully when urethral inflammation has excited seminal emissions. Dr. Baehr says, *digitaline* will cure the severest cases of involuntary seminal discharges, especially when there is great weakness and palpitation of the heart.

Other writers maintain that after all the voluntary causes have been removed, the involuntary effects that remain must be treated in accordance with the prominent symptoms, as in case of anæmia and debility and frequent pollutions, *ferrum pyrophosp.* and *china*, or in case of constipation and hemorrhoids *nux vomica* and *sulphur* administered alternately night and morning, or if strangury is a prominent symptom attendant on pollutions, and painful erections, *cantharis* 3d or 6th given three times a day before meals will generally cure SPERMATORRHŒA, which consists of an involuntary discharge of seminal fluid, when at stool or when urinating, or at other times from the slightest exciting causes, and especially when there are but feeble erections, or flaccidity of the penis ; and when there is great weakness of the back and spine *conium maculatum* may be given in the 6th decimal dilution three times a day before meals ; *calcis hypoph.* is also a remedy much esteemed. A small powder of the second decimal three times a day after meals has done well in many cases. *Hypophosphite of zinc* deserves a careful study. *Oxalate of cerium* is a valuable remedy for spermatorrhœa as borne out by clinical experience. A small powder of the 2d decimal may be given three times a day.

Ustilago madis in the hands of Dr.W. H. Burt cured a case of nocturnal emissions of long standing when other remedies had failed.

Cypripedium pubescens. Dr. E. M. Hale administered this remedy in a case of great nervous prostration and depression of spirits, and it seemed to impart new tone and vigor to the nervous system.

The remedies in general for spermatorrhœa include those prescribed for nocturnal emissions, as well as those known to act on the spinal centre, the most prominent of which are *conium mac.*, *digitaline*, *ferrum pyro.*, *nux vom.*, *plantago major*, *selenium*, and when there are great weakness, emaciation, dullness and depression, *china*, *phosphoric acid*, *sepia* and *sulphur*.

While taking remedies, great care is required to

avoid all medicinal articles of diet, all distilled and fermented liquors. The doses of the liquids where not mentioned are from 1 to 5 drops in water, to be repeated from one to four times in 24 hours.

By carefully studying the therapeutic hints given in the preceding pages, and making a practical application of the remedies pointed out, we confidently assert that no one need despair of deriving the most desirable benefits.

And, further, a confiding trust in Providence and a firm reliance upon the best appointed means will invigorate the whole system, dissipate fears, depression of spirits and physical weakness, restore happiness, and promote a certain return to manhood.

CHAPTER XIV.

THE APPLICATION OF ELECTRICITY IN THE TREATMENT OF SEMINAL WEAKNESS.

The effect of electricity upon the nervous system has received greater or less attention for several years. Of late it has been classed among the most effective remedial agents, and is applicable to those conditions which are dependant upon spinal irritation, and particularly upon atrophy of the cord. It is undoubtedly a useful remedy for torpid states of the nerves that convey the vis nervosa to certain parts, and as such it may be employed in the treatment of seminal weakness. But it is an unsuitable agent to be tampered with, and none but careful hands should undertake to administer it— and then according to explicit directions.

The Faraday Battery, which is an excellent apparatus for treating a variety of nervous difficulties, is not so desirable in the treatment of spermatorrhœa. The interrupted currents sometimes cause a succession of slight shocks, which, instead of increasing the vital activity of the seminal vesicles, act disastrously upon them. It is, therefore, important to have a suitable electrical apparatus, of light construction, that can be called into requisition when needed. The Voltaic, or Galvanic pile, is by far the most preferable for effective purposes. An ingenious pocket battery, invented by J. J. Geiger* is, in point of economy and superlative usefulness, the very best in use, and by the employment of it in many of the most inveterate cases of nocturnal emissions, as well as in the more obstinate cases of spermatorrhœa, it has been found, in most instances, to produce satisfactory results. But in order to promote its usefulness in these distressing maladies, it must be borne in mind that even the imponderable agents are powerless unless brought into certain relations in subserviency to the laws and conditions that govern them.

Therefore, in the treatment of these troubles, patients must deliberately, and with determination, abandon all exciting causes, avoid excesses in eating and drinking, and be always particular to shun couches of down or feathers, because the toleration of these might interfere with the use of the battery. It is also incumbent on patients to avoid all violent exercise or excessive physical exertion, and to keep the mind directed to cheerful and interesting topics—to keep good, social, and improving society, and to indulge in reading useful, improving, and entertaining books.

The Geiger Battery, with full directions for its use, can be obtained of Clindinning & Co., No. 35 South Clark street, Chicago, who are agents for the same. The mode of using them in the treatment of these affections is as follows: Once in two or three days, prepare the battery as directed, and place the *positive* electrode immediately back of the scrotum, and the *negative* part above the small of the back—use the *primary* current. It is said by some that this treatment is universally useful, and with one of Geiger's Batteries the current can be varied in force. At first it should be moderate and of brief duration, and afterwards, as the patient becomes accustomed to it, the force of the current may be increased, and the length of time it is employed may be increased to twenty minutes, if necessary.

*For sale by Clindinning & Co., 35 South Clark Street Chicago, Illinois.

www.ingramcontent.com/pod-product-compliance
Lightning Source LLC
Chambersburg PA
CBHW022014190326
41519CB00010B/1525